A GUIDE TO
FINDING & TRACKING
MINOR PLANETS AND COMETS

FOR EDUCATORS FOR AMATEURS

Designed and Written by Errol Jud Coder
Contributions by Brent A. Sorensen, MS.
Cover photo: Asteroid 433 Eros

Documents, photographs, and diagrams by Errol Coder
Research and some procedures curtsey the IAU Minor Planet Center
http://minorplanetcenter.net/

Printed in the United States
10 9 8 7 6 5 4 3 2 1

ISBN-13: 978-1466259492
ISBN-10: 1466259493

:Astronomy,
Comet & Asteroid Detection,
Tracking, Reporting and Detection

Want to send any results from your research to our
archive or order additional copies?
Email ***Errol.Coder@gmail.com*** for details.

NOTE Every effort has been taken to ensure that
all information in this book is correct and accurate.
Please contact us regarding any said errors.

Contents

Comet and Minor Planet Detection and Tracking Guide
For Education Programs and Amateurs

There are a number of Comets and Asteroids recognized as "Near Earth Objects." Those designated as "NEO" have an orbit that crosses the path of the Earth as it orbits around the sun. To aid in the awareness of these objects a number of observation programs have been developed using the help of professional and amateur astronomers. These important observers continue to study and identify these minor planets as they play "cosmic pinball" through out our neighborhood in space. Everyday new objects are identified and added to this growing list of potential hazards. Those classified as "NEO" are only a very small percentage of all the other comets and asteroids that exist in our local system. Identification and tracking of NEO and non-NEO Asteroids and Comets is infinitely the most important aspect of this form of research. A number of asteroids exist with in the

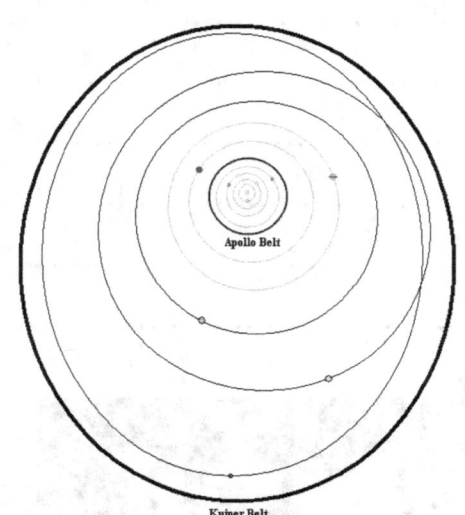

Apollo Asteroid field that orbits around the sun between Jupiter and Mars. Another field is found around Jupiter, the Trojan Asteroids at one of the three LaGrange points in its orbit. Other asteroids have been found in the Kuiper Belt which lies just past Neptune encircling the solar system. This field holds a good number of the asteroids and comets that end up being tugged into the Solar System by gravitational effects of nearby planets and sun. With in this field Pluto is sometimes considered the Grandfather of Comets as its structure resembles one. This area is said to hold the majority of asteroids created when the Solar system was formed. They were expelled by the other planets into an orbit around the outer planets. Beyond the Kuiper belt on the very edge of the Solar System is the Oort cloud which houses the vast majority of the Comets that have yet to be drawn into the solar system. While we have detected enough of these objects to have a hefty list, there are a vast number of others that have yet to be identified.

The guide is designed for just this reason. It was developed to provide aspiring observers a method they could apply to conduct their own searches. Ultimately, any new Minor Planets that are detected and identified will provide a greater understanding of the world around us, and allow us to know more about the threats with in our solar system. If we can broaden the knowledge of detection by informing amateur astronomers as to how this can be done, the effectiveness of the efforts will increase substantially.

Designed by
Errol Coder
Planetary Sciences/Observational Astronomy, Undergraduate Student
eratimus@gmail.com

Scientific Advisor
Brent Sorensen, MS
Associate Professor of Physics and Astronomy – Southern Utah University

Basic Observation Equipment

Telescope (reflector or refractor

For Photographic Imaging: Single Lens Reflex (SLR) Camera /w a T-Adaptor compatible to the telescope.

For Digital Imaging: You will need to know the focal length of your telescope and the physical size of your DSLR or CCD pixels to calculate the pixel scale. Your setup should be such that the pixel scale is no greater than 2 arc seconds/pixel (preferably) or 3 arc seconds/pixel (at worst). In practice, your optimal pixel scale is something that you will have to determine for yourself, taking into consideration the capabilities of your telescope and DSLR/CCD and the seeing at your site. If your pixel scale is much larger than the values quoted above, then the quality of the astrometry will suffer. If your pixel scale is too low for your local setup, then the signal-to-noise of the images may be low as each image is spread over a large number of pixels.

You will also need a computer to store the images and software to perform the reductions and refinements. Most astrometry software packages use the Guide Star Catalogue. You may also choose to use Starry Night. To aid you in image corrections photography programs such as Photoshop are highly recommended, as you can correct for contrast, brightness, color, and other needed modifications.

An accurate clock/watch set to Coordinated Universal Time (UTC) is a must and this must be checked regularly (as a minimum, at the start of each observing session) against a reliable standard such as the Naval Observatory Master clock that can be found at: http://tycho.usno.navy.mil/frtime.html

Access to e-mail is also important, both for reporting observations to and receiving designations from the Minor Planet Center.

Internet access is also important to submitting observations to the Minor Planet Center and to access the New Ephemeris Generation website to generate data for tracking your objects.

A Planisphere is recommended to aid you in quickly locating objects. Each Planisphere is designed to show how the sky would appear at a given latitude on the Earth. Make sure the one you use is designed for your latitude.

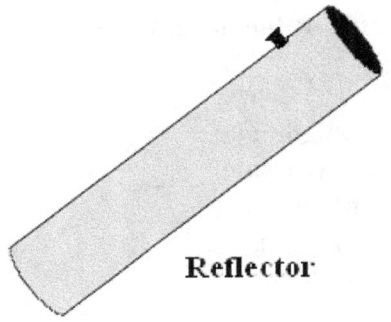

Reflector

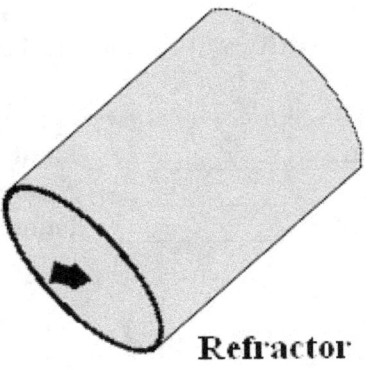

Refractor

Step one: Determining Equipment Limiting Magnitude

An objects magnitude is its apparent brightness/illumination that the observer sees. An object may be closer, yet may appear dimmer then another object that is farther away yet appears to be brighter. An object that is brighter then another may actually be farther a way your viewing, but may be quite larger then the closer/dimmer object or releases a larger amount of energy and light. Most often an Asteroid or Comet would appear fainter then the background stars, as they do not generate their own light, and have less surface area to reflect the light from the sun, which is the only way you can see them in the dark void. It is important to know the limiting magnitude of your equipment. Knowing the faintest object that can be observed will assist you in understanding the apparent magnitude of the faintest asteroid or comet that you may also see.

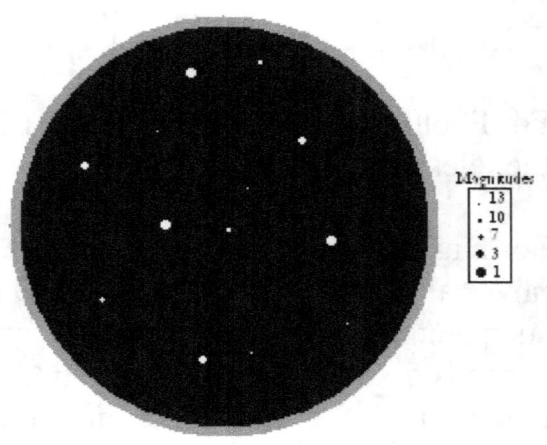

The simplest way to evaluate limiting magnitude is to select a commonly known constellation group. Once selected, start your observations at the location. Locate an object with in the area of your field of view that appears to be the faintest. This object would appear to be the smallest and to be emitting less light then the others. Using a star chart, try to locate the object. Using the star charts magnitude table, try and compare how the object appears on the map to the chart. Most often there is a table of larger dots shrinking to smaller dots, going from a lower number (larger dot/brightest) to a higher number (bigger dot/faintest). As your viewing the field of view at magnification, not all the stars you see may appear on the chart you're using. Use the best judgment in determining what the magnitude of the faintest star is. If the star your viewing appears fainter then the faintest star that is on the map, you can usually estimate the magnitude.

In this example, using a magnitude reference chart from our imaginary star map, it appears that the faintest observable object has a magnitude of 13. It would appear that the faintest asteroid or comet that you may end up being able to see, using this example could be no fainter then 13 magnitude.

Focus on three different regions and determine the Limiting Magnitude of the observed field.

COMPLETE – *Photocopy this page for additional uses*

Constellation: _____ Faintest Object: _____ Magnitude: _____
Constellation: _____ Faintest Object: _____ Magnitude: _____
Constellation: _____ Faintest Object: _____ Magnitude: _____
Average: _____

Step two: Determining the Equipments Field of View (FOV)

In order to become more accustomed to the telescope you're using, a simple exercise to determine the *angular diameter* of the field of view of the telescope is needed. The field of view is that portion of the sky visible through the telescope. The field of view is different for each eyepiece. **Higher magnification** results in a **smaller field of view**. Knowing the field of view of the telescope for each eyepiece gives you a reference to estimate the size of the celestial objects you are observing. It also allows you to understand the area which you are viewing when compared to a star map. For small objects like asteroids and comets, or for more precise measurements, the method of transit times is to be preferred. You will use the method of transit times to measure the angular diameter of the field of view of the telescope. Since the celestial sphere (the sky) appears to rotate in one day, the stars drift through the field of a stationary telescope. Since the rotation of the sky is well known (360° in 24 hours), we can use that fact to measure the field of view of the telescope by simply timing how long it takes for a star to drift through the field of view.

Example:

Suppose for example that a star takes 2 minutes to drift across the diameter of the field of view. Then   we can readily compute the field of view (FOV):

FOV = 2 min *(1 hour/60 min) * (360°/24 hour) = 0.50°

OR 30' (minutes of arc)

This is approximately the apparent diameter of the Moon. In this telescope/eyepiece combination, the Moon would barely fit in the field of view.

The above example is strictly valid only if we use a star located on the **celestial equator** (declination = 0°). Stars located away from the celestial equator also make one turn in the sky in 24 hours, but their apparent motion takes place on smaller circles and their angular speed across the sky is smaller. An extreme example is the pole star, Polaris. It is located very close to the North Celestial Pole and if you point your telescope at it with the drive turned off, you will see that it hardly moves at all. This is exactly the same situation as we have on Earth. Someone standing at the North Pole would make one turn in one day, but wouldn't be moving at all. At the same time, people on the equator are the ones who are moving the fastest due to the Earth's rotation. We can correct for this effect in our field of view measurement by introducing the cosine of the declination (latitude in our Earth example) of the star you are using. The field of view is correct given by:

FOV = (transit time in hours) * cos(declination) * (360°/24 hours)

Note the following conversions for a star located on the celestial equator:

Transit time	Angle
24 hours	360°
1 hour	15°
1 minute	15'
1 second	15"

Recall that 1° (degree) = 60' (minutes of arc) = 3600" (seconds of arc)

Procedure

- Using star charts, identify a naked-eye star within 10° of the celestial equator.

- With the telescope drive ON, acquire the star in the telescope, using one size eyepiece available for that particular telescope. Put the star in the center of the field of view.
- Turn the drive OFF. Using the RA slow motion knob (East-West motion), move the star to just outside of the field of view so that it will drift back in on its own. The star should go through the center of the field, i.e. travel along the diameter of the field of view.
- Using a stopwatch, time how long it takes for the star to drift across the field. Make the same measurement 3 times, re-centering the star each time.
- Repeat using another size eyepiece used for the telescope your calculating the FOV.
- In the measurement record below, note the time, the name and declination of the star (read off the star chart to the nearest degree)

When you are all done, compute the average transit times and the field of view in degrees and in arc minutes, rounded off to 0.01°/1', for each eyepiece. Write these numbers clearly so you can refer to them easily for the rest of your observations.

You may choose to make observations with the use of a camera, while either mounted on your telescope or on a separate tripod. When you are viewing the FOV with a telescope, you're seeing it as a round field and you need only to measure the drift times across one edge to the other. But, when you're viewing with a camera, the field of view will be a rectangle. Instead of recording the drift of a star crossing from one edge to another, you will need to measure the drift times across the horizontal and vertical view of the camera.

This is done just like a telescope. Starting with the longest edge, horizontal view set the camera view so a star is visible on one edge of the field of view. Set it up so the star will drift from one edge of the rectangle to the other. Next, rotate the camera so the shortest edge, the vertical view. Set the star so it will pass through the middle of the field of view.

Once you have the transit times for both sides of the view, as with the telescope, calculate the dimensions of each edge of the camera view.

COMPLETE – *Photocopy this page for additional uses*

Take 5 FOV measurements from the same object and derive at an average duration for the object crossing the FOV of your telescope:

Eyepiece (mm)

_____

Object name: _____

Time 1: _____
Time 2: _____
Time 3: _____
Average: _____
Calc. FOV:

_____

Eyepiece (mm)

_____

Object name: _____

Time 1: _____
Time 2: _____
Time 3: _____
Average: _____
Calc. FOV:

_____

Camera Lens:

_____

Horizontal Edge

Time 1: _____
Time 2: _____
Time 3: _____
Average: _____
Calc. FOV: _____

Object name:

_____

Vertical Edge

Time 1: _____
Time 2: _____
Time 3: _____
Average: _____
Calc. FOV: _____

Field of View (FOV)

Horizontal   **Vertical**

_____ X _____

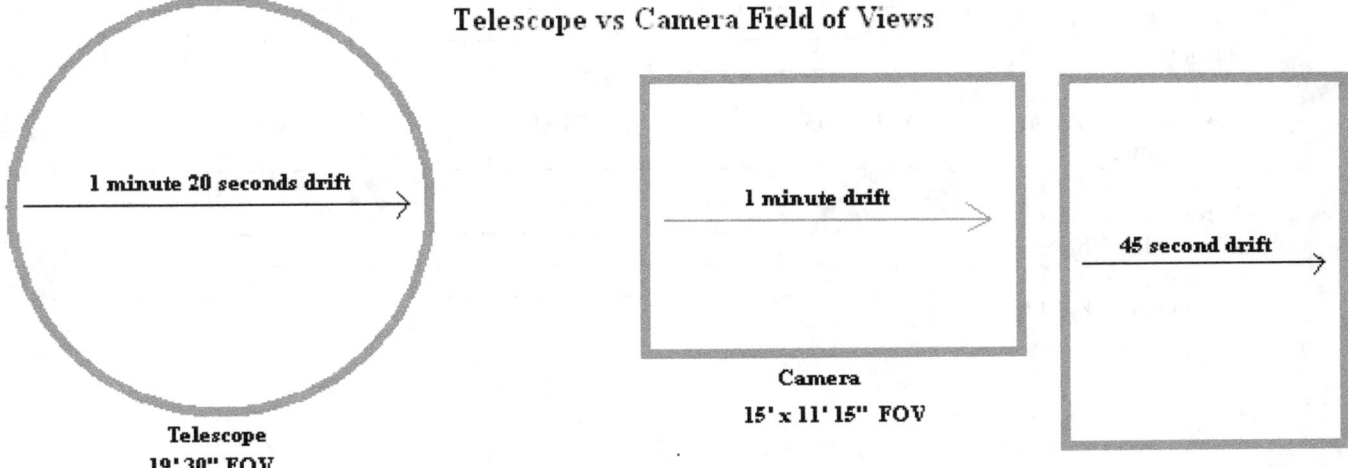

Telescope vs Camera Field of Views

1 minute 20 seconds drift

1 minute drift

45 second drift

Telescope
19' 30" FOV

Camera
15' x 11' 15" FOV

Step three: Designating Observation Locations

A high number of comets and asteroids orbit around the sun and generally appear along the ecliptic, the same orbital plane that the planets tend to follow. There is a greater chance of spotting an object along the ecliptic then anywhere else. Select 4 bright objects in constellations +/- 20° declination from the ecliptic. These 4 objects will act as the central guide stars for the 4 observation areas. Using a star map, identify the 5 brightest objects surrounding the 4 guide stars that are with in your telescopes field of view. You can identify these objects by using a star map. During each observation of the particular area you will center your telescope on the guide star, making two observations per period and sketches of the visible objects with in the field of view. Each observation of the area, it is suggested to try and identify all visible objects. You may use a Digital Camera mounted on the telescope as an alternative to sketching.

COMPLETE – *Photocopy this page for additional uses*

Observation Area 1:	Designation	RA/DEC	Constellation
Center Object:			
Surrounding Object 1:			
Surrounding Object 2:			
Surrounding Object 3:			
Surrounding Object 4:			
Surrounding Object 5:			

Observation Area 2:	Designation	RA/DEC	Constellation
Center Object:			
Surrounding Object 1:			
Surrounding Object 2:			
Surrounding Object 3:			
Surrounding Object 4:			
Surrounding Object 5:			

Observation Area 3:	Designation	RA/DEC	Constellation
Center Object:			
Surrounding Object 1:			
Surrounding Object 2:			
Surrounding Object 3:			
Surrounding Object 4:			
Surrounding Object 5:			

Observation Area 4:	Designation	RA/DEC	Constellation
Center Object:			
Surrounding Object 1:			
Surrounding Object 2:			
Surrounding Object 3:			
Surrounding Object 4:			
Surrounding Object 5:			

Step four: Minor Planet Reference Observation

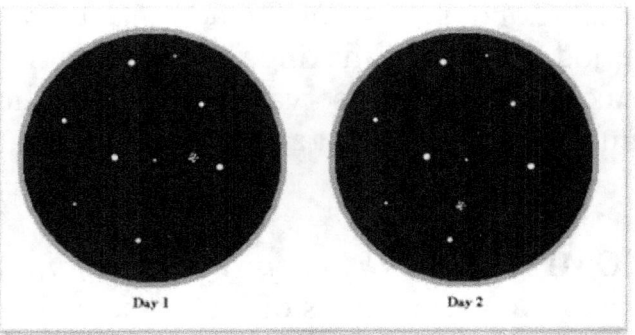

As stars, galaxies, nebulae and objects outside of our solar system are quite a distance away, when observed they would not appear to change position in respect to each other. If you were to observe a group of stars one night, and return the next, the objects would be positioned at the same location in respect to their neighboring stars. While they may be oriented different due to the Earth's orbit, they will basically appear not to move when compared to the others. Asteroids and Comets are different. As they are with in the solar system, and are much closer, they would appear to move across the background sky just like a planet would. If you were to observe a field of stars and return the next day, any object or point of light that appears to be out of place when compared to the previous observation is most likely one of these objects, having moved as it travels through the solar system. Unless they are quite close, and you're using a high powered telescope, they would still appear as points of light. But, they would be points of light moving across the background sky when observed over a period of days or weeks.

   Example observations conducted over a 2 day period. The object shifts from one position to another in respect to the background sky.
   Using any resource, locate an asteroid or comet that will be present in the night sky during the time of year that your observations will be conducted that is visible with in your limiting magnitude.

COMPLETE – *Photocopy this page for additional uses*
Make two observations per night twice a week of the comet or asteroid. Use the Observation log when conducting observations. After the two nights of observation, submit the data to generate Ephemeris tracking data. (See step 8)
Object Type _____   **Name** _____   **Constellation** _____

Observation Summary	App. Mag.	RA	DEC
Night 1 Obs 1			
Night 1 Obs 2			
Night 2 Obs 1			
Night 2 Obs 2			

Step five: Making Observations

Use the Observation and Photography Logs to keep track of your observations. Make sure to observe and photograph the designated areas as close to the same time each observation period, i.e. if your making your observation of one area at 9:00 PM one night, make sure to make your future observations of that particular area as close to 9:00 PM each and every time your observe that area.

COMPLETE – *Photocopy this page for additional uses*

Make two observations of each observation area at least 3 times a week separated by a day between observations. If you are limited on your available time, twice a week, separated by at least 3 days between observations would be acceptable.

		Week 1	Week 2	Week 3	Week 4	Week 5	Week 6	Week 7	Week 8
Area 1	Day 1	❑ Obs 1	❑ Obs 1	❑ Obs 1	❑ Obs 1	❑ Obs 1	❑ Obs 1	❑ Obs 1	❑ Obs 1
		❑ Obs 2	❑ Obs 2	❑ Obs 2	❑ Obs 2	❑ Obs 2	❑ Obs 2	❑ Obs 2	❑ Obs 2
	Day 2	❑ Obs 1	❑ Obs 1	❑ Obs 1	❑ Obs 1	❑ Obs 1	❑ Obs 1	❑ Obs 1	❑ Obs 1
		❑ Obs 2	❑ Obs 2	❑ Obs 2	❑ Obs 2	❑ Obs 2	❑ Obs 2	❑ Obs 2	❑ Obs 2
	Day 3	❑ Obs 1	❑ Obs 1	❑ Obs 1	❑ Obs 1	❑ Obs 1	❑ Obs 1	❑ Obs 1	❑ Obs 1
		❑ Obs 2	❑ Obs 2	❑ Obs 2	❑ Obs 2	❑ Obs 2	❑ Obs 2	❑ Obs 2	❑ Obs 2
Area 2	Day 1	❑ Obs 1	❑ Obs 1	❑ Obs 1	❑ Obs 1	❑ Obs 1	❑ Obs 1	❑ Obs 1	❑ Obs 1
		❑ Obs 2	❑ Obs 2	❑ Obs 2	❑ Obs 2	❑ Obs 2	❑ Obs 2	❑ Obs 2	❑ Obs 2
	Day 2	❑ Obs 1	❑ Obs 1	❑ Obs 1	❑ Obs 1	❑ Obs 1	❑ Obs 1	❑ Obs 1	❑ Obs 1
		❑ Obs 2	❑ Obs 2	❑ Obs 2	❑ Obs 2	❑ Obs 2	❑ Obs 2	❑ Obs 2	❑ Obs 2
	Day 3	❑ Obs 1	❑ Obs 1	❑ Obs 1	❑ Obs 1	❑ Obs 1	❑ Obs 1	❑ Obs 1	❑ Obs 1
		❑ Obs 2	❑ Obs 2	❑ Obs 2	❑ Obs 2	❑ Obs 2	❑ Obs 2	❑ Obs 2	❑ Obs 2
Area 3	Day 1	❑ Obs 1	❑ Obs 1	❑ Obs 1	❑ Obs 1	❑ Obs 1	❑ Obs 1	❑ Obs 1	❑ Obs 1
		❑ Obs 2	❑ Obs 2	❑ Obs 2	❑ Obs 2	❑ Obs 2	❑ Obs 2	❑ Obs 2	❑ Obs 2
	Day 2	❑ Obs 1	❑ Obs 1	❑ Obs 1	❑ Obs 1	❑ Obs 1	❑ Obs 1	❑ Obs 1	❑ Obs 1
		❑ Obs 2	❑ Obs 2	❑ Obs 2	❑ Obs 2	❑ Obs 2	❑ Obs 2	❑ Obs 2	❑ Obs 2
	Day 3	❑ Obs 1	❑ Obs 1	❑ Obs 1	❑ Obs 1	❑ Obs 1	❑ Obs 1	❑ Obs 1	❑ Obs 1
		❑ Obs 2	❑ Obs 2	❑ Obs 2	❑ Obs 2	❑ Obs 2	❑ Obs 2	❑ Obs 2	❑ Obs 2
Area 4	Day 1	❑ Obs 1	❑ Obs 1	❑ Obs 1	❑ Obs 1	❑ Obs 1	❑ Obs 1	❑ Obs 1	❑ Obs 1
		❑ Obs 2	❑ Obs 2	❑ Obs 2	❑ Obs 2	❑ Obs 2	❑ Obs 2	❑ Obs 2	❑ Obs 2
	Day 2	❑ Obs 1	❑ Obs 1	❑ Obs 1	❑ Obs 1	❑ Obs 1	❑ Obs 1	❑ Obs 1	❑ Obs 1
		❑ Obs 2	❑ Obs 2	❑ Obs 2	❑ Obs 2	❑ Obs 2	❑ Obs 2	❑ Obs 2	❑ Obs 2
	Day 3	❑ Obs 1	❑ Obs 1	❑ Obs 1	❑ Obs 1	❑ Obs 1	❑ Obs 1	❑ Obs 1	❑ Obs 1
		❑ Obs 2	❑ Obs 2	❑ Obs 2	❑ Obs 2	❑ Obs 2	❑ Obs 2	❑ Obs 2	❑ Obs 2

Step six: Observation Analysis

Whether you are sketching your observation, using a Digital, Film, or CCD Camera, it is a simple process to analyze your observations. The best way to identify possible comets or asteroids is to compare past images of the areas with recent observations of the same area. This will allow you to observer whether or not there is an object that appears to be moving in respect to the known stationary objects in your field of view. If you do observe a point of light that appears to be moving in respect to the rest, then chances are you have hit the jack pot. But, remember you're looking at a small slice of the sky, so it may be a while of constant observations and image analysis before you locate an object. The reason you're observing along the ecliptic, is that just like the planets, most asteroids and comets also follow this path. So by making your observations along this path, it increases your chances of locating an object.

If you're using a CCD or Digital Camera, you can immediately transfer your images to your computer and do your comparisons right away. It will also allow you to make image adjustments.

If you're using a film camera, you can still develop your images as soon as you wish. You will not need to process your first observation image until have you have completed a series of other observations. Once you have made a few observations, you can not develop your pictures and make your image comparisons. If you have the ability to develop your own negatives and prints, it is best to do so this way if you're using film. You can process your film rolls into black and white. You would then have the ability to make your comparisons using strictly the negatives. If you find acceptable objects that seem to move as an asteroid or comet would on your negative, you can then make only prints of those images you want.

Observation Examples

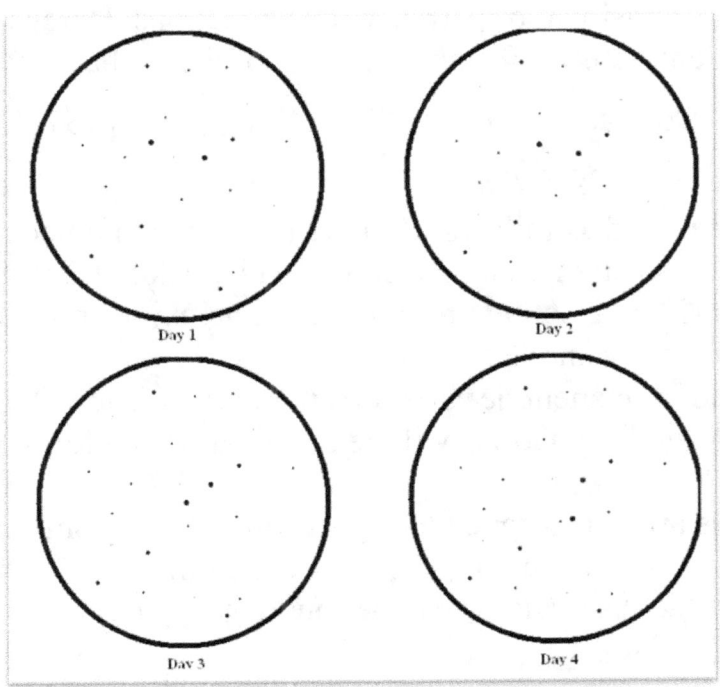

Step seven: Object Identification, Submission & Verification

Once you have observed an object that is unknown to you, or you are following-up on a previous object, there is a specific format you need to follow to have your observations accepted by the Minor Planet Center.

Prior to submitting the object you will need to assign a temporary designation. New provisional designations are assigned to newly-reported objects that cannot be identified with a known numbers, multi-opposition unnumbered or recently-discovered one-opposition (with or without a general orbit) minor planet.

New designations are assigned upon the receipt of observations on two nights. The two nights should be fairly close together, certainly within a week of each other.

If there are a number of observers involved at a particular site and assignment of credit for the discovery of particular objects is important, ensure that the observer-assigned temporary designations reflect the names of the discoverers. For example, at a particular site there are three observers—Coder, Welch, and Sorensen. Objects discovered by Coder alone are reported with temporary designations beginning Co (e.g. Co0001) using no more then 6 total characters. Objects discovered by Coder and Welch should be designated jointly by designations beginning CW or CoWe (e.g., CW0001 or CoWe01), designations beginning CoSo indicate objects discovered by Coder and Sorensen.

It is preferred that discoveries are made by a single individual, although discoveries by pairs of discoverers are accepted. Claims for discoveries of specific objects by three or more discoverers are treated as site discoveries, where no individuals are named as the discoverer. If you're new object (observed on two or more nights) can be matched to an earlier un-designated one-night stand, that one-night stand does not have priority and you will receive the credit for the discovery. If your new object can be matched to an earlier designated one-night stand (generally this will be a re-designation from a published night object), your new object will have priority.

If after you have submitted an observation of a potential new object and you get a reply that you have "new" objects you, you will receive a list matching your temporary designations to official provisional or permanent designations. Here is a (fictitious) sample assumed to have been sent Feb. 1999, showing most of the probable forms:

Co0001   *(03244*   *CoWe01*   *J99A18T*   *CoSo02*   *(J81U78A*   *Co0004*

(Co00003)   *Co0003*   *(J99A08J*

This may be interpreted as follows: Co0001 is numbered object (3244); CoWe01 is a new object 1999 AT_{18} that is credited to Coder and Welch; CoSo02 is the known unnumbered object 1981 UA_{78}; Co0003 and Co0004 refer to the same object, now designated 1999 AH_8, which is a recent discovery by another team.

In short, provisional and permanent designations not preface with '(' are your discoveries. Provisional and permanent designations will be in the packaged form, as used on the observation record.

New designations are not assigned to objects observed on only one night, although you may receive designations if such objects can be identified with already-known objects. Instructions for submitting observations can be found at http://cfa-www.harvard.edu/iau/mpc/html

Step eight: Generating Ephemeris Data for Tracking
(referenced from the Minor Planet Center Astrometry Guide)

If you need to generate ephemeris tracking data for an object that is already identified then you need to visit the Ephemeris Generator located @ http://ssd/jpl.nasa.gov/cgi-bin/eph

If you need to generate ephemeris tracking data for a new unidentified object, so you may continue to conduct observations until you receive an identification notice from the Minor Planet Center, you will need to use the information you have reported out in your object observation records. There needs to be a minimum of two days of observation data collected to submit an ephemeris request. The records are formatted to resemble the format needed to be submitted at the New Object Ephemeris Generator website located @ http://www.cfa.harvard.edu/iau/MPEph/NewObjEphems.html

As a note, if you do not know any of the information requested on the observation record, simply leave the field blank. When you enter the data in the ephemeris submission field, leave the fields that are blank, blank unless the legend below instructs otherwise.

Format for Observation Logs, Object Reporting and New Object Ephemeris Generator Submissions

NNNNNPPPPPPPANNYYYY MM DD.ddddd HH MM SS.dd sDD MM SS.d   MM.M B   OOO

For Minor Planets (leave blank any info you do not know)

Columns	Format	Use
1 – 5	NNNNN	Minor planet number
6 – 12	PPPPPP	Provisional or temporary designation
13	A	Discovery asterisk

Minor planet numbers and provisional designations are official designations assigned by the Minor Planet Center. Temporary designations are designations, preferably no more than six (6) characters long (the absolute maximum is seven (7) characters), assigned by the observer for new or unidentified objects. Temporary designations must consist of alphanumeric characters only: don not include spaces. If the object you're viewing is new and unidentified, leave the Minor planet number field blank.

For Comets (leave blank any info you do not know)

Columns	Format	Use
1 – 4	NNNN	Periodic comet number
5	N	Letter indicating type of orbit
6 – 12	PPPPP	Provisional or temporary designation
13		Not used, must be blank

Periodic comet numbers and provisional designations are official designations assigned by, respectively, the Minor Planet Center and Central Bureau for Astronomical Telegrams. Temporary designations are designations, up to six (6) characters long, assigned by the observer for new or unidentified objects. In practice, temporary designations on comet observations will be very rare. If the object you're viewing is new and unidentified, leave the Periodic comet number field blank.

For Minor Planets and Comets (leave blank any info you do not know)

Columns	Format	Use
1 4	N	Note 1
15	N	Note 2
16 - 32	YYYY MM	Date of observation Year/Month/Day

	DD.ddddd	
33 - 44	HH MM SS.dd	Observed RA (Hour/Minutes/Seconds
45 – 56	sDD MM SS.d	Observed Decl. (+/-Hour/Minutes/Seconds
57 - 65		Must be blank
66 – 71	MM.M B	Observed magnitude and band
72 - 77		Must be blank
78-80	OOO	Observatory code

Detailed Notes:

Minor Planets

NUMBER

Columns 1-5 contain a zero-padded, right-justified number—e.g., an observation of (1) would be given as 00001, an observation of (3202) would be 03202. If there is no number these columns must be blank. Six-digit numbers are to be stored in a packed form (A000 = 100000), in order to be consistent with the format specifier earlier stated.

PROVISIONAL/TEMPORARY DESIGNATIONS

Columns 6-12 contain the provisional designation or the temporary designation. The provisional designation is stored in a 7-character packed form.

DISCOVERY ASTERISK

Discover observations for new (or unidentified) objects should contain '*' in column 13. Only one asterisked observation per object is expected.

Comets

PERIODIC COMET NUMBER

Periodic comets that have been observed at more than once return are assigned numbers.

ORBIT TYPE

Column 5 contains 'C' for long –period comet, 'P' for a short-period comet. 'D' for a 'defunct' comet, 'X' for an uncertain comet or 'A' for a minor planet given cometary designation.

PROVISIONAL DESIGNATION

Columns 6-12 contain a packed version of the provisional designation. The first two digits of the year are packed into a single character in column 6 (I = 18, J = 19, K = 20). Columns 7-8 contain the last two digits of the year. Column 9 contains the half-month letter. Columns 10-11 contain the order with in the half-month. Column 12 will be normally '0', except for split comets, when the fragment designation is stored there as a lower-case letter.

Examples:

1995 A1   = J95A010

1994 P1-B = J94P01b   refers to fragment B of 1994 P1

1994 P1   = J94P010   refers to the whole comet 1994 P1

Columns 6-12 may contain a minor-planet provisional designation. In such a situation column 12 will contain a capital letter.

Comets and Minor Planets

NOTE 1

This column contains an alphabetical publishable note or a numeric or non-alphanumeric character program code.

NOTE 2

This column serves two purposes. For those observations which have been converted to the J2000.0 system by rotating B1950.0 coordinates this column contains 'A', to indicate that the value has been adjusted. For those observations reduced in the J2000.0 system this column is used to indicate how the observation was made. The following codes will be used"

P   Photographic (default if column is blank)
E   Encoder
C   CCD/DSLR
T   Meridian or transit circle
M   Micrometer
V/v   "Roving Observer" observation
R/r   Radar observation
S/s   Satellite observation
c   Corrected-without-republication CCD/DSLR observation
o   Occulation-derived observations

In addition, there is 'X' which is used only for already-filed observations. It was given originally only to discovery observations that were approximate or semi-accurate and that had accurate measures corresponding to the time of discovery: this has been extended to other replaced discovery observations. Observations marked 'X' are to be suppressed in residual blocks. They are retained so that an original record exists of a discovery.

DATE OF OBSERVATIONS

Columns 16-32 contain the date and UTC time of the mid-point of observation. If the observation refers to one end of a trailed image, then the time observation will be either the start time of the exposure or the finish time of the exposure. The format "YYYY MM DD.dddddd", with the decimal day of observation normally being given to a precision of 0.00001 days. Where such precision is justified, there is the option of recording times to 0.000001 days.

OBSERVED RA (J2000.0)

Columns 33-44 contain the observed J2000.0 right ascension. The format is "HH MM SS.ddd", with the seconds of R.A. normally being given to a precision of 0.01s. There is the option of recording the right ascension to 0.001s, where such precision is justified.

OBSERVED DECL (J2000.0)

Columns 45-56 contain the observed J2000.0 declination. The format is "sDD MM SS.dd" (with "s" being the sign), with the seconds Decl. normally being given to a precision of 0.01". There is the option of recording the declination to 0".01, where such a precision is justified.

OBSERVED MAGNITUDE AND BAND

the observed magnitude (normally to a precision of 0.1 mag.) and the band is which the measurement was made. The observed magnitude can be given to 0.01 mag., where such a precision is justified. The default magnitude scale is photographic, although magnitudes may be given in V- or R-band, for example. For comets, the magnitude must be specified as being nuclear, N, or total, T.

The current list of acceptable bands is: B (default if band is not indicated), V, R, I, J, C, W, U, g, r, I and z. Non-recognized magnitude bands will cause observations to be rejected. Addition of new recognized bands requires knowledge of a standard correction to convert a magnitude in that band to V.

OBSERVATORY CODE

Observatory codes are stored in columns 78-80. Lists of observatory codes are published from time to time in the Minor Planet Circulars. Note that new observatory codes are assigned only upon receipt of acceptable astrometric observations.

Step nine: Measuring Parallax (Advanced Procedure)

One of the few reliable methods astronomers may use to determine the distances to objects in space is called **parallax**. It is based on trigonometry.

There are special mathematical tools one can use to determine the lengths of the sides of a right triangle. In the figure below, suppose that we want to know the distance L, which represents the distance between the observer and the minor planet.

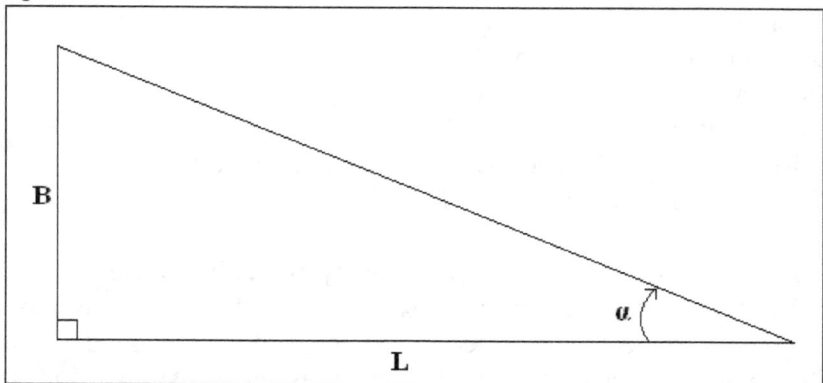

If we can measure the side **B**, and the angle **alpha**, then we can use the equation

$$L = \frac{B}{\tan(\text{alpha})}$$

To calculate **L**.

We can use the same method for a situation which there are two such triangles, back-to-back:

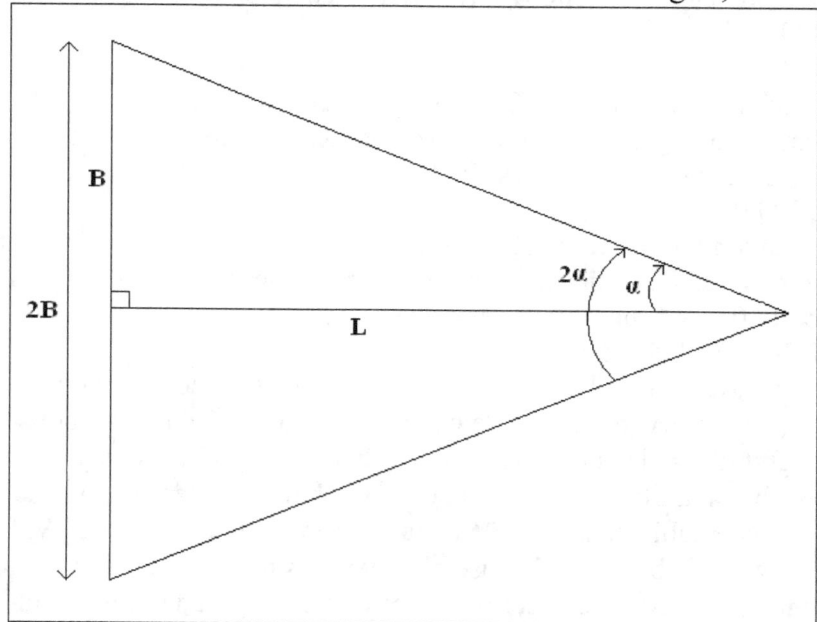

We can still figure out the distance **L** with trigonometry. Once again, we calculate.

$$L = \frac{B}{\tan(\text{alpha})}$$

But we could also describe this as

$$L = \frac{\frac{1}{2}\,(2B)}{\tan\left[\,\frac{1}{2}\,(2+\text{alpha})\,\right]}$$

Or

$$L = \frac{\text{half of the total baseline}}{\tan\left[\,\text{half the total angle}\,\right]}$$

Measuring angles is difficult. It's not an easy thing to determine exactly in which direction you are looking at a particular moment. Remember that astronomers, sitting on the Earth's surface, are constantly being rotated around the Earth's axis (once every 24 hours), and carried along with the Earth around the Sun (once every 365 days). It's very hard for us to keep track of the absolute direction of our telescopes when we look at stars, planets, or even asteroids.

There are times when it helps to use **parallax** to determine *relative* angles. Relative measurements, in astronomy and in all of physics, are often much easier than absolute ones. Parallax is simply the apparent shift of a nearby object *relative to objects behind it* as one moves from one end of a baseline to another: to see it, hold your arm outstretched in front of you, and hold one finger straight up. Hold your arm steady. Close one eye, and look past your finger at some distant object. Now close the OTHER eye and look past your finger again. You should see that it appears to move, relative to the distant object.

Astronomers can do the same thing. If they can observe the same nearby object from two different places, they may detect a difference in its appearance relative to the background stars.

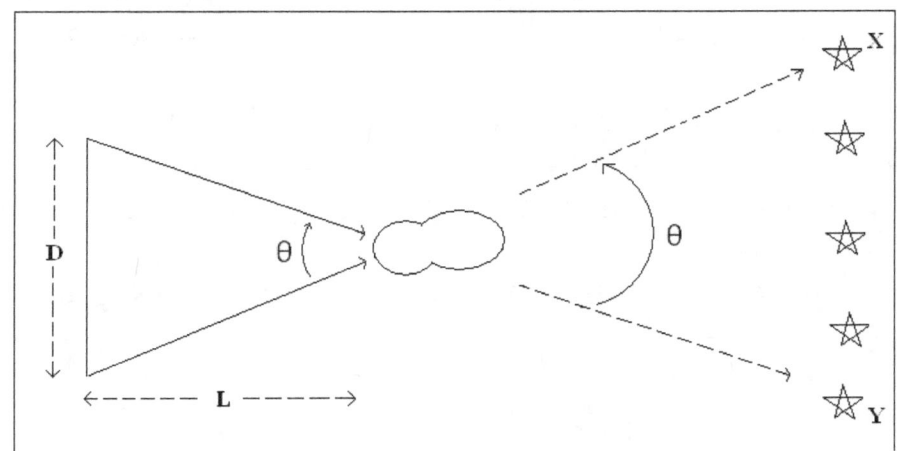

Suppose that two astronomers separated by a distance **D** observe an asteroid simultaneously. One sees the asteroid next to star X, and the other sees it next to star Y. The angle between the two observed locations for the asteroid refers to its angular separation. To determine this, by using a star chart compare the Right Ascension for the position of the asteroid for one observer to its location on a star map for the other observer. For example: If one observer calculates the asteroid as having a right ascension of **15.3°** and the other observer at some distance away from the other observer determines that the Asteroid has a right ascension measured at 23.8 degrees, then **23.8° – 15.3°** = an angular separation of **8.5°**. So θ for both the angular separation of the two observed locations of the asteroid by the background sky and the angle of the two observers to the asteroid equals **8.5°**. So, in this case θ = **8.5°**. If the astronomers know the angle between the two observed locations of the asteroid, and they know the distance between their observatories, they can calculate the distance **L** to the asteroid:

$$L = \frac{\text{half of the total baseline}}{\tan\left[\text{half the total angle}\right]}$$

$$= \frac{D/2}{\tan\left[\theta/2\right]}$$

> *This method only works when the background objects – the stars – are much, much farther away than the asteroid. Only then is the angle between the stars, as seen from the Earth, the same as the angle between the stars, as seen from the asteroid. Fortunately, in real life, the stars really are much, much farther away than any body in our solar system, so we can use this method.*

Using parallax to determine the distance to a Planet

The same method would be used if you detect an asteroid and wish to calculate its distance. The use of a planet is only to allow practice and familiarity of the process.

This process will require the participation of two observers a distance from one another of at least 1,000 mi/15893 km apart. Select a prominently visible planet in the night sky. Make arrangements for both observers to simultaneously conduct observations of the planet at the same time on the same nights. Make three observations over 3 separate days with no more then one day apart. Make sure that both individuals make the observations at the exact same time.

Object _____   D = _____ km

	Observer 1 Lat/Long				Observer 2 Lat/Long			
	_____				_____			
Day 1	Time (UTC)	Date (UTC)	Guide RA/DEC	Θ	Time (UTC)	Date (UTC)	Guide RA/DEC	Θ
	_____	_____		__	_____	_____	_____	__
Day 2	Time (UTC)	Date (UTC)	Guide RA/DEC	Θ	Time (UTC)	Date (UTC)	Guide RA/DEC	Θ
	_____	_____		__	_____	_____	_____	__
Day 3	Time (UTC)	Date (UTC)	Guide RA/DEC	Θ	Time (UTC)	Date (UTC)	Guide RA/DEC	Θ
	_____	_____		__	_____	_____	_____	__

Day 1

$$= \frac{D/2}{\tan[\,theta/2\,]}$$

L = ---------------- =
km

Day 2

$$= \frac{D/2}{\tan[\,theta/2\,]}$$

L = ---------------- =
km

Day 3

$$= \frac{D/2}{\tan[\,theta/2\,]}$$

L = ---------------- =
km

References

Internet Websites
1. Ephemeris Generator – http://ssd/jpl.nasa.gov/cgi-bin/eph
2. Near Earth Object website – http://neo.jpl.nasa.gov/neo/tools.html
3. Small Body Identification – http://ssd.jpl.nasa.gov/cgi-bin/sb_search
4. Small Body Orbital Elements –
5. SkyMorph Moving Target Detection – http://skys.gsfc.nasa.gov/cgi-bin/skymorph/mobs.pl
6. Minor Planet Center - http://cfa-www.harvard.edu/iau/mpc/html

Books and Reference Guides
1. NORTON's 2000.0 Star Atlas and Reference Handbook
2. Burnham's Celestial Handbook, Vol I - III

Computer Programs
1. STARRY NIGHT Pro

Bibliography

Norton, Arthur P., Ridpath, Ian, ed. Norton's 2000.0 Star Atlas and Reference Handbook. Essex, England: Longman Group UK Limited, 1996.

Parallax in theory and practice. 9 May 1999. 3 Sept. 2008. <http://www.tass-survey.org/classes/phys236/parallax/parallax.html>

Minor Planet Center. 1 Jan 2008. Minor Planet Center 2 Sept. 2008. <http://cfa-www.harvard.edu/iau/mpc/html>

The Planetary Society Gene Shoemaker Grant. 4 Sept. 2008. The Planetary Society. <www.planetary.org>

Sorensen, Brent. Associate Professor of Physics and Astronomy. Southern Utah University. 2003, 2004, 2005, 2006.

Cotts, Laura. Physics Lab Instructor. Southern Utah University. 2003, 2004.

Glossary

Angular diameter the apparent size of a celestial object, usually expressed in degrees, minutes and seconds of arc.

Angular distance the apparent distance between two objects on the celestial sphere, such as two stars, usually expressed in degrees, minute and seconds of arc.

Arc (measure of) angles on the celestial sphere are measured in degrees, minutes and seconds of arc. The terms *arc minute* and *arc second* are used to distinguish these measures from unites of time. There are 60 arc minutes in a degree, and 60 arc seconds in an arc minute.

Asteroid a small celestial body orbiting the sun, which is composed of iron, nickel, ice and other material created at the formation of the solar system.

Astrometry the branch of astronomy concerned with the precise measurement of the positions of objects on the celestial sphere.

Ephemeris a table of the predicted positions of a celestial object such as the Moon, the Sun, or an asteroid.

Ephemeris a table of the predicted positions of a celestial object such as the Moon, the Sun, or an asteroid.

Guide star a star oriented in the middle of the field of view within the same field as an object under study, but which lies at a different distance and hence has no connection. The guide star is used as a central point to allow the observer to return to the area of observation and to focus on the same location.

Meteorite a chunk of rock or iron from space that reaches the surface of the Earth or any other body. Large meteorites can produce craters when they hit the ground. Most meteorites are thought to be chips from asteroids, but some fragile stony meteorites called carbonaceous chondrites may come from the nuclei of comets.

Planisphere a circular map with a rotating mask that can be turned to show the stars as they appear from a given latitude at any time on any date.

Position angle the relative position of one object with respect to another, such as the two components of a double star or the position of a star around the Moon's limbs at an occultation. Position angle is measured in degrees from north via eat, south, and west. On the celestial sphere, east is the direction towards the eastern horizon.

www.ingramcontent.com/pod-product-compliance
Lightning Source LLC
Chambersburg PA
CBHW081818280526

45789CB00008B/3146